# STEM CONCEPTS

LUNA OWL
PRESS

$\epsilon F = ma$

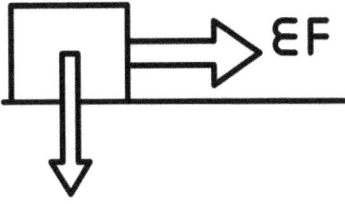

$\epsilon F$

$E_K = \frac{1}{2}mv^2$

# PHYSICS

$E = m.c^2$

$\Sigma F = M.a$

# GRAVITY

Gravity is the force that pulls everything down!

Without gravity, we would float away into space!

# FORCE
## (PUSH AND PULL)

**A force is a push or a pull that makes something move.**

FUN FACT: You use force when you kick a ball, open a door, or pull a wagon!

# FRICTION

Friction is a force that happens when two things rub against each other. It can make things slow down or stop moving.

## FUN FACT
You rub your hands together to warm them up because friction creates heat!

# MAGNETISM

**FUN FACT:**
Magnetism is a force that pulls certain metals.

**FUN FACT:**
Earth is a giant magnet!

**FUN FACT:**
Magnets have a north pole and a south pole,

# SIMPLE MACHINES

**Fun Fact:**
Simple machines make work easier!

**Fun Fact:**
A lever helps lift heavy things with less effort!

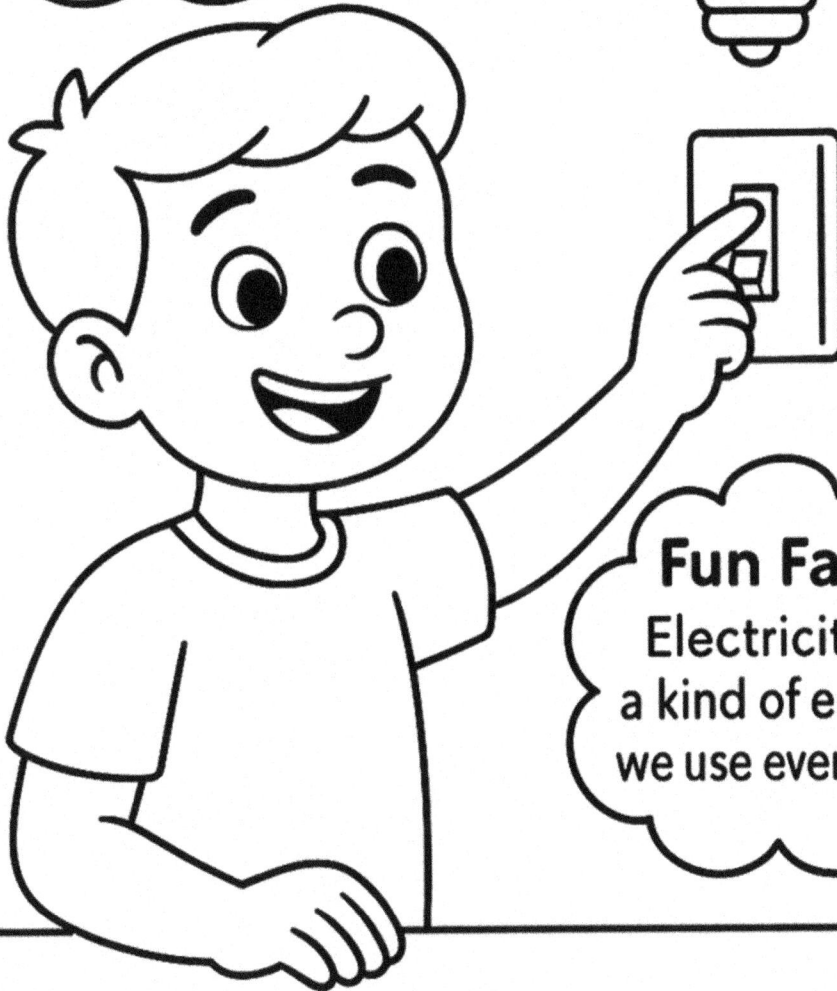

# MATH

$$E = mc^2$$

$$F = ma$$

Math helps scientists understand how the world moves!

We use math to measure force, energy, speed, and more

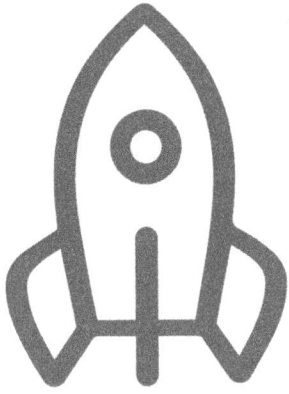

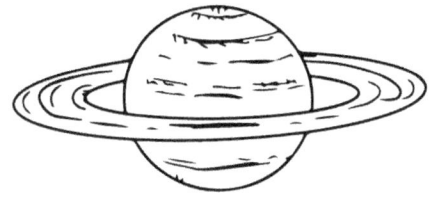

# SPACE AND ASTRONOMY

# SOLAR SYSTEM

The Solar System has 8 planets that orbit the Sun!

The Sun is a giant star at the center!

# ASTRONAUT

Astronauts float because there's almost no gravity in space!

Space suits protect astronaults fromm cold and heat!

Space suits protect astronauts from the cold and heat!

# STARS AND CONSTELLATIONS

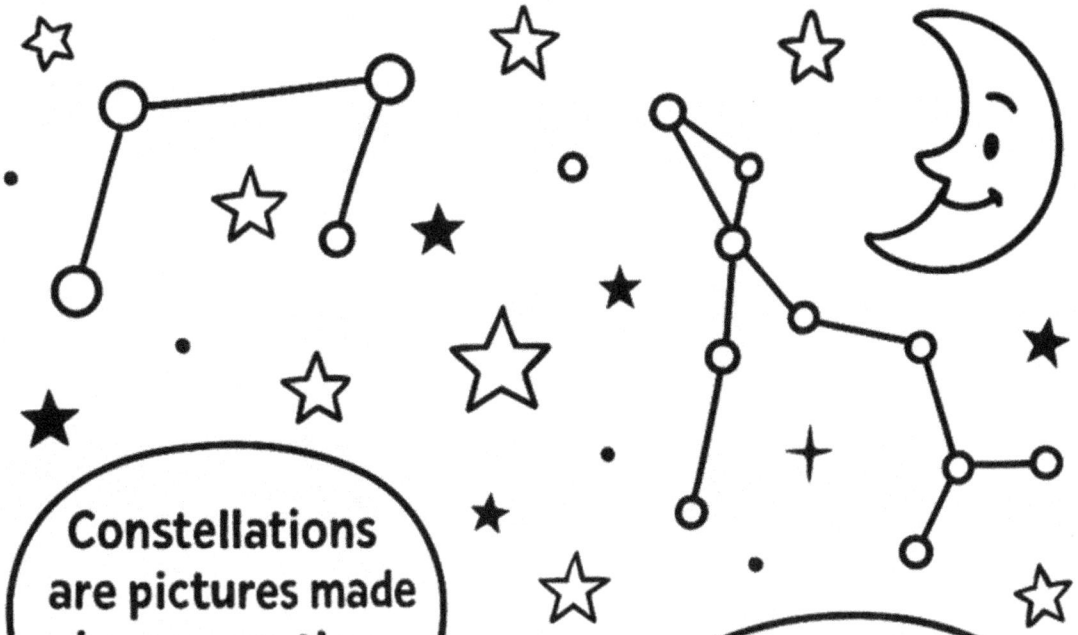

Constellations are pictures made by connecting stars!

People have used constellations to find their way for thousands of years!

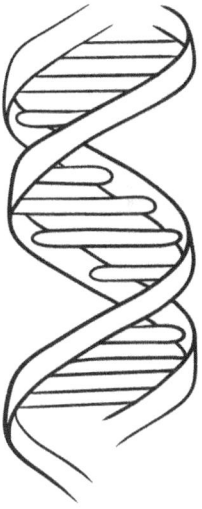

# BIOLOGY AND LIFE SCIENCE

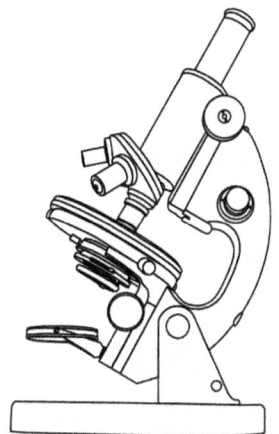

# PLANT LIFE CYCLE

Plants grow from tiny seeds!

Plants need water, sunlight, and soil to grow!

# LIFE CYCLE OF A BUTTERFLY

**Fun Facts:**
Butterfliies
start as tiny
eggs!

**Fun Facts:**
A caterpillar changes
inside a chrysalis
before becoming
a butterfly!

# FOOD CHAIN

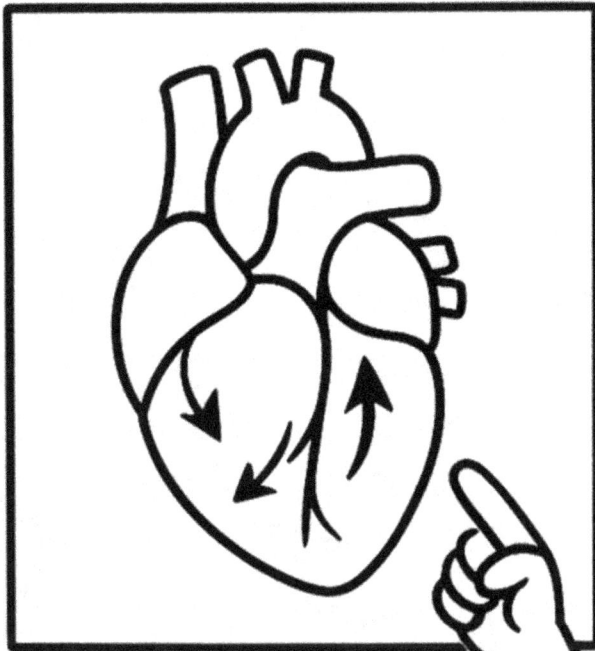

# LUNGS

**Fun Fact!**
Your lungs fill with air when you breathe in!

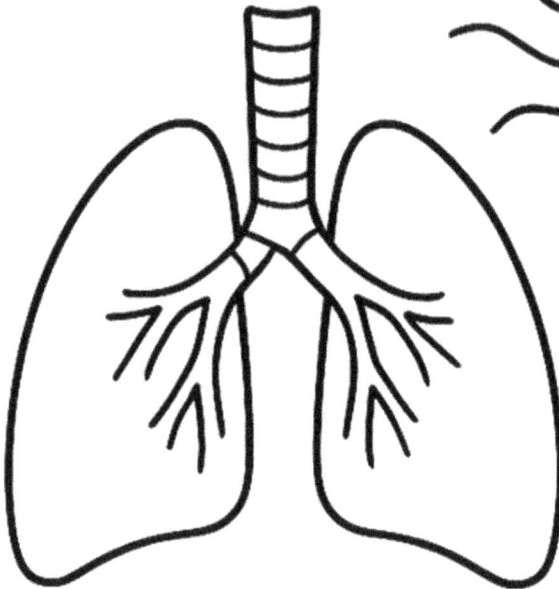

**Fun Fact!**
Oxygen from the air gives your body energy

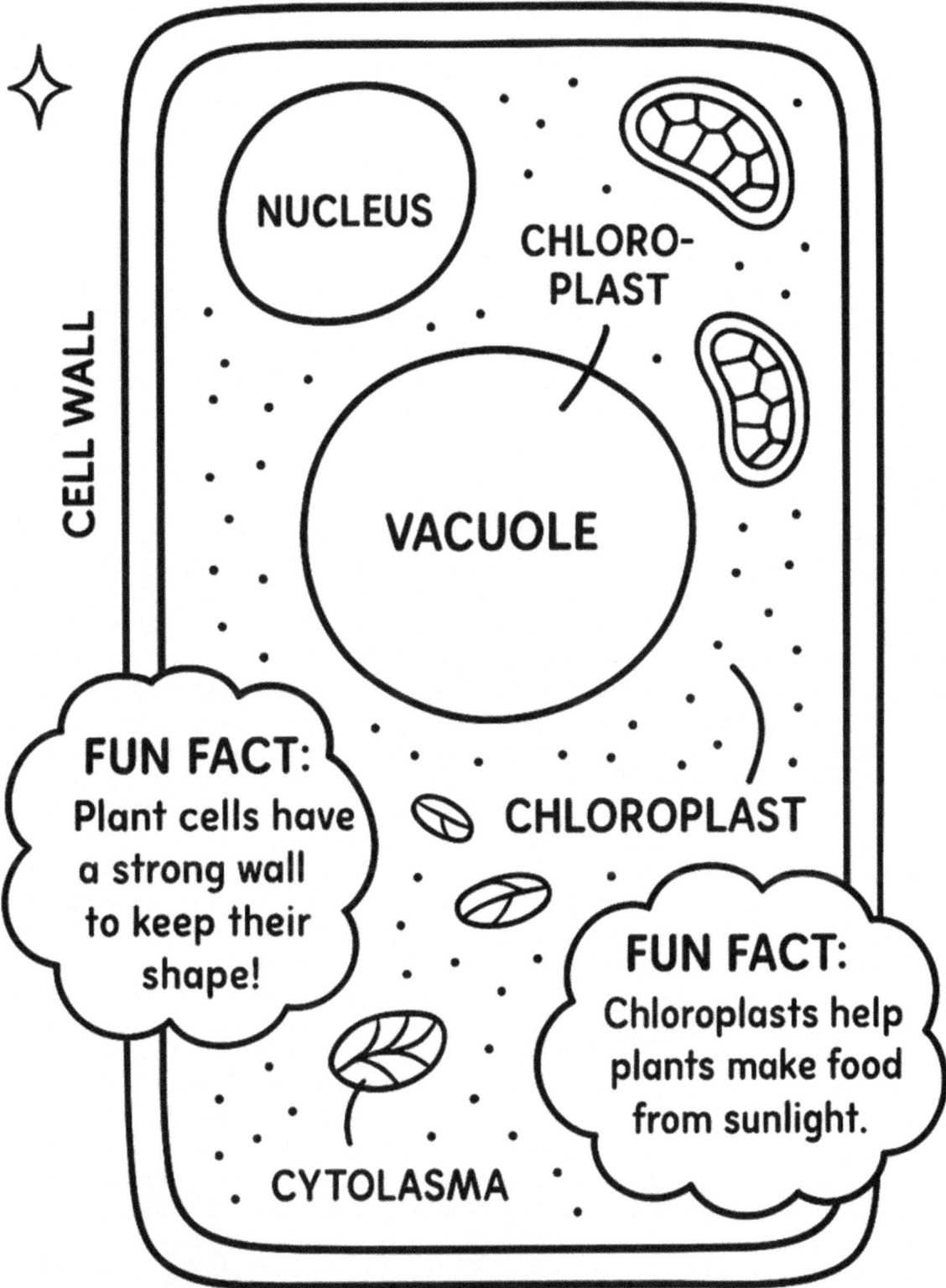

# PLANT CELL

NUCLEUS

CHLORO-PLAST

CELL WALL

VACUOLE

**FUN FACT:** Plant cells have a strong wall to keep their shape!

CHLOROPLAST

**FUN FACT:** Chloroplasts help plants make food from sunlight.

CYTOLASMA

# ANIMAL CELL

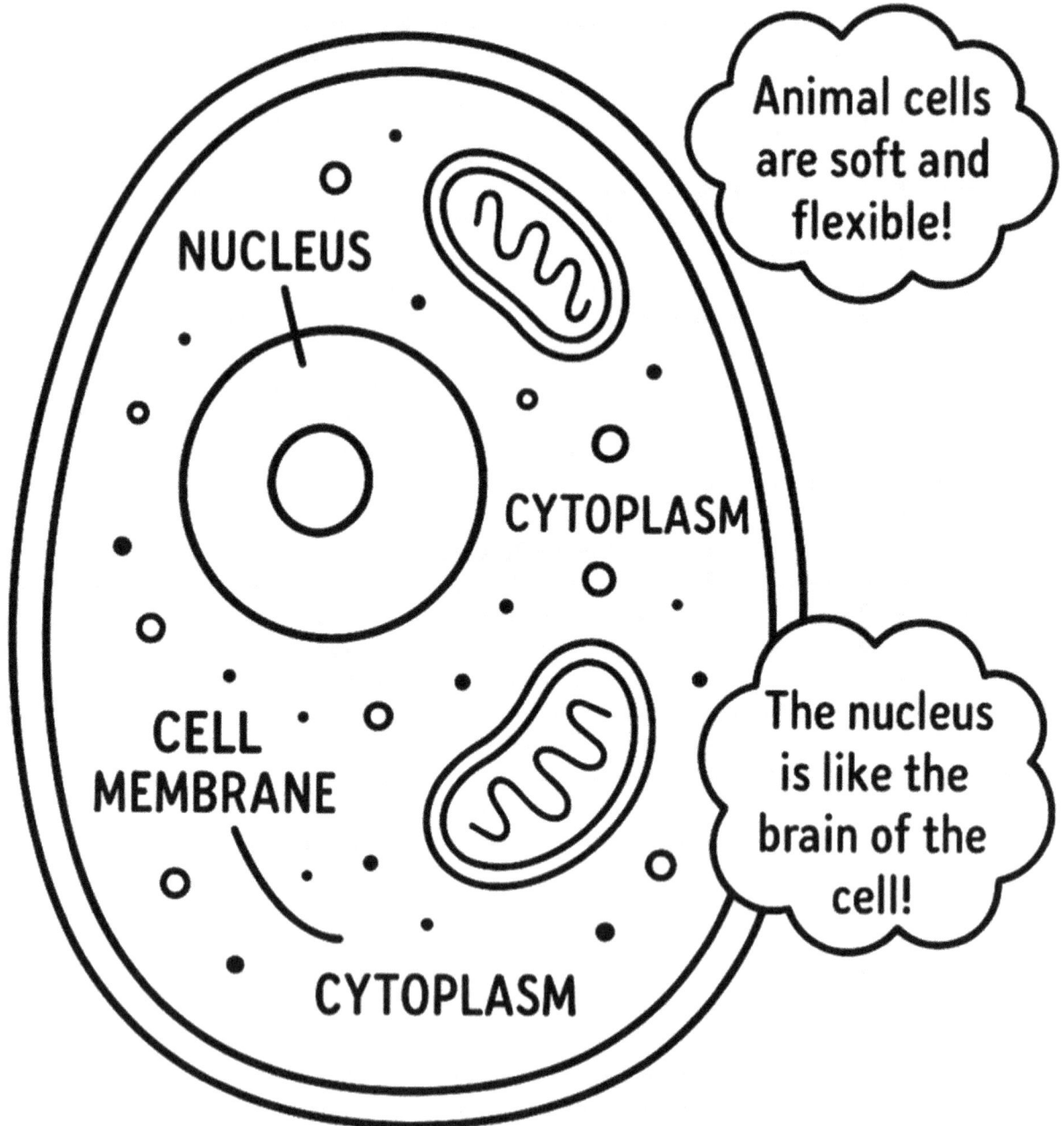

NUCLEUS

Animal cells are soft and flexible!

CYTOPLASM

CELL MEMBRANE

The nucleus is like the brain of the cell!

CYTOPLASM

Animal cells are soft and flexible!
The nucleus is like the brain of the cell.

# ANIMAL HABITATS

DESERT

OCEAN

FOREST

**FUN FACT!**
Animals live
In habitats
that give them
food, water,
and shelter

**FUN FACT!**
Different animals
live in deserts,
oceans, forests,
and ice

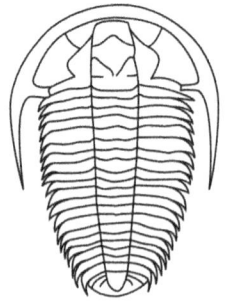

# EARTH SCIENCE

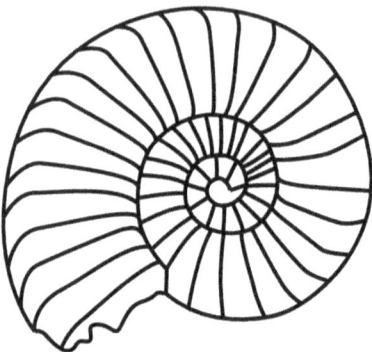

# VOLCANOES

Volcanoes form when melted rock called magma bursts out!

Some volcanoes are tall mountains, others are underwater!

# TYPES OF ROCKS

Igneous rocks form from cooled lava!

IGNEOUS

SEDIMENTARY

METAMORPHIC

Sedimentary rocks are made of layers pressed together!

# FOSSILS

Fossils are the remains of plants and animals from long ago!

Scientists study fossils to learn about Earth's history.

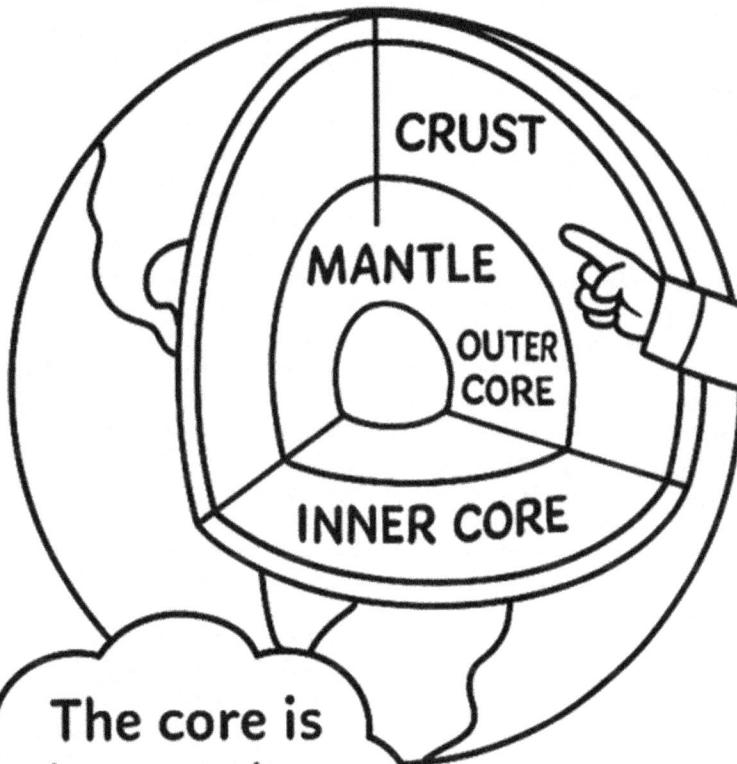

# WEATHER PATTERNS

**Fun Fact!**
Weather is how the air feels outside!

**Fun Fact!**
Rain, snow, and wind are all types of weather.

**Fun Fact:**
Rain, snow, and wind are all types of weather.

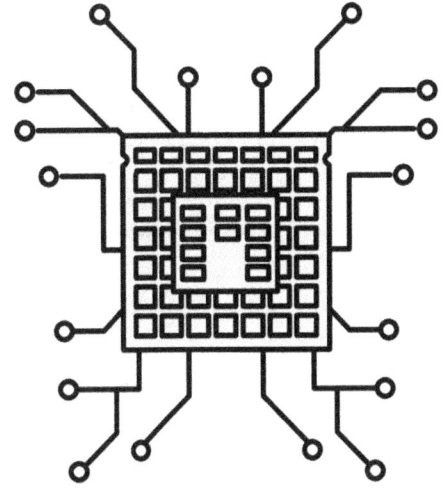

# TECHNOLOGY AND ENGINEERING

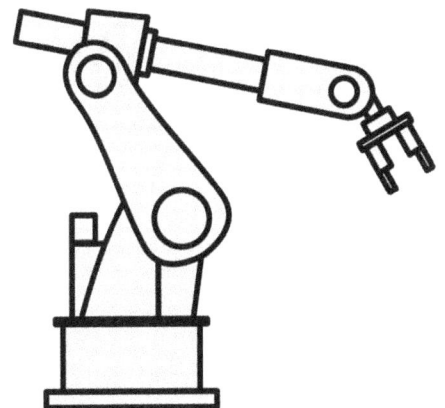

# 3D PRINTING

**Fun Fact:**
3D printers create real objects layer by layer!

**Fun Fact**
You can print tays, tools, and even houses.

**Fun Fact**
You can print **toys, tools** and even houses.

# CIRCUIT BOARD

**Fun Fact**
Circuit boards connect electricity to make things work!

**Fun Fact**
They are inside computers, phones, and robots

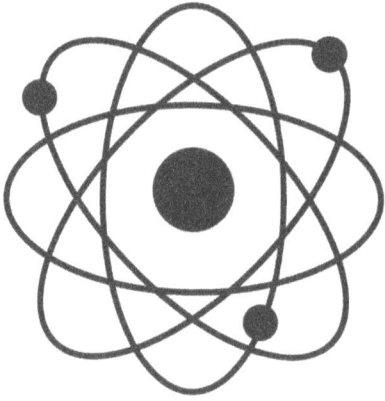

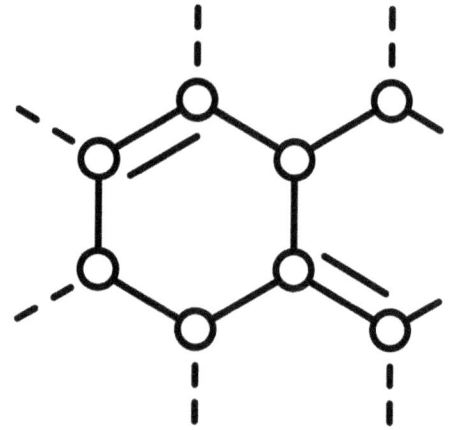

# CHEMISTRY

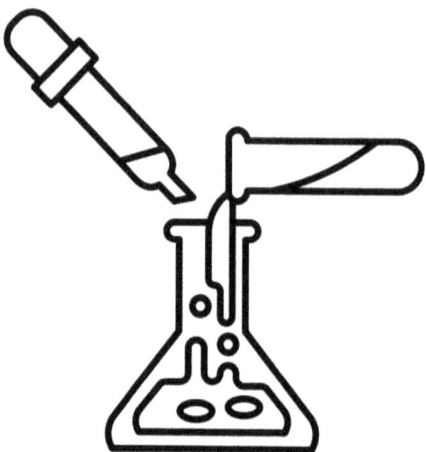

# PERIODIC TABLE

The periodic table shows all the elements in the universe!

Each element is like a building block of everything around us!

# STATES OF MATTER

**SOLID**

**LIQUID**

**GAS**

Fun Fact
Matter can be
a solid, liquid,
or gas!

Heat can change
matter from one
state to another!

# CHEMICAL REACTION

**FUN FACT**

A chemical reaction happens when things mix and change!

**FUN FACT**

Bubbles, color changes, and fizz are signs of a reaction.

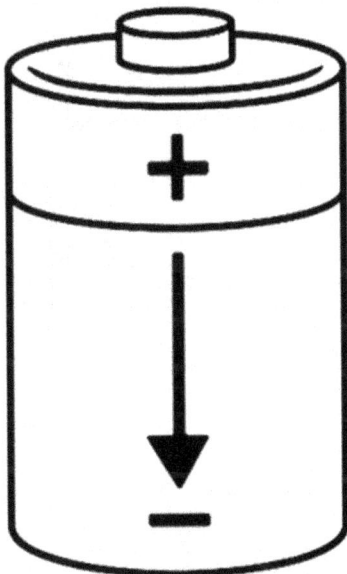

# CHEMICAL BONDING

Chemical bonds hold atoms together!

Molecules are groups of atoms stuck together!

# CERTIFICATE
## OF COMPLETION

This award is proudly presented to:

for exploring the amazing world of STEM—
science, technology, engineering, and math—
through creativity, curiosity, and imagination!

Date _____

Signed _____

www.ingramcontent.com/pod-product-compliance
Lightning Source LLC
Chambersburg PA
CBHW051801200326
41597CB00025B/4649